Solar PV Off-Grid Power:

How to Build Solar PV Energy Systems for Stand Alone LED Lighting, Cameras, Electronics, and Remote Communication Power Systems

by Christopher Kinkaid

Solardyne.com

Published by Solardyne, LLC
Portland, Oregon

ISBN-13: 978-1500473372
ISBN-10: 1500473375

Table of Contents

Preface

Solar energy is a formidable resource. Solar Electric Power systems, based on PV panels, make effective power supplies for your off-grid electricity needs. The sun distributes over 1,000 watts per square meter at peak, and is the natural power supply for life on Earth. The Sun, can also be your Power Supply.

The best kept industrial secret, is that we don't need to burn fossil fuels for industrial power. Solar PV panels, true 21st century tools, can provide daily energy production which can be used directly, or stored for later use, on demand, to power your remote electric loads, onsite, with no pollution, or fuel costs.

This Book is written to be a resource in building your own Solar PV supply for remote Cameras, LED lighting systems, Communication, Sensors, and remote Cabin, and Home Power systems, with Solar PV Power system examples.

The Solar Energy resource varies with time of day, season, and local climate. Solar PV panels, sized properly, produce reliable, and predictable energy production, despite daily variations, when calculated properly for each month.

Tap into PV Panels to charge battery banks for reliable DC, and, with inverters, AC power on demand. Remote site power supplies, designed, and installed properly, offer real power for running a variety of electronic, motor, and large draw devices.

Use this Book to match your Energy Load, with the Energy Production sized to match your electric loads for remote Solar PV Power. System examples range from 30 Watt Solar PV Power supplies for cameras, electronics, and sensors, to 4,000 Watt Home Power Systems.

About the Book

This Book is written as a step-by-step guide to defining your solar power projects "vital statistics," and choosing the right equipment to get the job done. If you have a specific solar powered project in mind, then visit the Solar PV Powered System Examples List located at the Quick Guide in Chapter Nine.

The **Quick Guide** takes you to a specific Solar Water Pumping System. Sample systems, provide Part Lists, so you can configure your own systems, or match your electric load to the closest system included in this Book, from 30 watt systems, to 4,000 Watts.

Chapter 1 covers the Solar Energy Resource, and the "Big Picture" issues which define the best way to put Solar PV energy to work.

In **Chapter 2** outlines the Step-By-Step process to define your system for powering your loads. From Cameras, Electronics, Water Pumping, and Treatment, on demand, and Remote Home Power systems, solar PV Power is effective as a Power Supply.

Chapter 3 discusses the aspects, and issues with Solar PV Panels. Size, Power Rating, Energy Rating, Mounting options. Solar Peak Hours.

Chapter 4 discusses, and lists, Solar PV Power Systems from 30 watts to 120 Watts. Systems include Charge Controllers, Batteries, Mounting Hardware, and Inverters for your AC loads. Includes System Examples.

Chapter 5 covers Solar PV Power Systems from 135 watts to 360 Watts. Popular as Power Supplies for LED area lighting, LED sign lighting, Water Treatment, Cameras, Sensors, Communication Platforms, and Remote Cabins. System Examples included.

Chapter 6 examines Solar PV Power Systems from 500 watts to 1,500 watts. Solar arrays, power conditioning, Battery Banks, Battery Enclosures, Inverters.

Chapter 7 looks at Solar PV Power Systems from 2,000 watts to 4,000 watts. Battery Bank Voltages, Power Panels, Power conditioning, Fusing, Safety Disconnects, Grounding.

Chapter 8 includes Solar PV Power Systems **Quick Guide** to reference Example Solar PV Systems Power, and Energy Ratings.

About the Author

Christopher Kinkaid

Christopher (Toby) Kinkaid, originally from Portland, Oregon, is the founder of **Solardyne.com**, **SolarQuote.com**, and **AlgaeToday.com**, and has worked in clean energy, and publishing technology for over three decades.

Kinkaid, is the inventor of the "**Helyx**" Vertical Axis Wind Generator, the "**Mariposa**" Non-imaging solar concentrator PV module (continuous operation at Sandia National Laboratory since 1994), the **Solar Demultiplexer** optical solar concentrating lens (Dr. James/Sandia National Laboratory 1991), and the inventor of the original "Solar Power Pack" (Mother Earth News, "Littlest Utility" June/July, 2001).

Kinkaid, has lectured on clean energy technology around the world including "APEC", Bangkok, Thailand, 2003, "Energy Solutions World", Tokyo, Japan, 2003, The International Biomass Conference

(IBC), 2010, Minneapolis, MN, and the Algal Biomass Organization (ABO) Conference, 2010, Phoenix, AZ.

Christopher (Toby) Kinkaid, has appeared in interviews on KOIN TV, KGW TV, and "Sustainable Today" produced in Oregon, and has served on the board of directors for the National Hydrogen Association, in Washington D.C., 1993, the Japan Satellite Communications Company (JCNET), Fukuoka, Japan, 1994-95, and Algaedyne Corporation, St. Paul, MN, 2010-2013.

Kinkaid, presently serves as CEO of Solardyne, LLC in Portland, Oregon, where he continues his work in Solar, Wind, and Biomass Technology applications, research, and development.

Introduction

Solar PV power systems, are an effective choice in remote power supplies. Electronic devices require reliable, and rugged power systems in remote locations, and at sites in extreme conditions, such as high altitude, tropics, or deserts.

Power your remote electronics, cameras, sensors, water pumping, water treatment, cabin, and home power with the range of Sample Solar PV Power systems included in this Book. Includes Example Systems, from 30 watts to 4,000 Watts.

Solar Photovoltaic (PV) panels convert solar energy into "hard currency" electricity to do work. Charge Batteries for power on demand 24/7. If your site is remote, Solar PV Power systems are a cost-effective way to produce power onsite, and gives you independence, no toxicity, and no fuel costs.

Fossil fuel Generators are noisy, pollute, and present ever increasing fuel costs. In remote locations, the transportation of fuels, often exceeds, the costs of the fuel. Solar PV power systems have matured, over the decades, and state-of-the-art PV panels are robust, reliable, and with no moving parts, long lived. Most Solar PV panels offer 25 year performance warranties.

Using Sealed, Maintenance-free batteries, stand alone solar PV power systems can be sized properly to power your load reliably, with the lowest cost.

The following Chapters are written to take you from general discussion, to specific Solar PV Power Systems you can use directly, or adapt to your specific requirements.

To Calculate the best Solar PV Power System for your remote site, work the problem "backwards." Start with the Energy Demand of your site, and match with the Energy Production of the Solar PV Power System.

This Book is written as a resource to determining the best Sizing, and Hardware for powering your remote load with reliability, no pollution, and no fuel costs.

Chapter One - Solar Power the Big Picture

Solar power is a force of nature, delivering over 1,000 watts, per Square meter at peak. This energy can be converted with high efficiency, using photovoltaic (PV) panels, into electricity to be used directly, or stored in batteries for power on demand.

In this book, we'll break down the questions you'll need to ask to define your system requirements. Then, match those requirements to the appropriate Solar Off-Grid power supply type and specification to get the job done.

The Power in the Sun is enormous, and can be easily tapped to power your electronics. Power your remote cameras, sensors, LED Lighting Systems, and remote homes using Photovoltaic (PV) panels to Power the front end your system.

The Solar resource:

Natural sunlight contains many wavelengths (colors) of light, which can be used, separately, for different purposes. Short wavelengths present in solar energy, like Ultra-violet (UV), are ideal for water sterilizing, and treatment. Short UV photons have high energy, and are capable of causing photo-chemical reactions.

Visible wavelengths, from Violet, Indigo, Blue, Green, Yellow, Orange, and Red, getting progressively longer in wavelength, are excellent for Solar Photovoltaic (PV) electricity production.

The Longest wavelengths present in sunlight, the Infra-Red (IR), are ideal for thermal applications, such as heating Air, or Water. However, for Photovoltaics (PV) power converters, the ideal wavelengths, depend on the Photovoltaic material.

Plants, in Photosynthesis (the source of Earth's Oxygen, and Base-Food Chain Nutrition) use selected wavelengths in the Visible spectrum from 400 to 700 nm. Photovoltaics, use all wavelengths in Sunlight which contain Energy Greater Than the "Band gap" of the solar cell material for electricity production.

The "Energy" in a photon "Increases" the shorter the wavelength (Einstein's Photoelectric Effect, 1905). Long wavelength photons, like Red, and IR have lower energies, respectively.

Most solar PV panels are made from Silicon. Silicon Solar PV panels have a "Band gap" of 1.1 electron-volts (eV). This means Silicon, the material, responds to wavelengths of light, 1,100 nm, (Red) and "shorter" to produce electricity. Infra-Red (IR) wavelengths in Sunlight Longer than 1,100 nm wavelength is too weak to produce electrons (electricity).

PV Solar Cells made from Silicon, in the old days, we're called "Red" cells, because they started to generated electricity at Red wavelengths of light.

Other PV materials, such as Gallium Arsenide (GaAs) have a much higher "Band gap," than Silicon. GaAs has a band gap of about 1.45 eV. GaAs based solar cells are called "blue" cells because they convert the shorter wavelengths present in sunlight.

Note: Red light does not have enough energy at 1.1 eV to produce any electricity in a GaAs solar cell. Only photons with "Energy" above the "Band-gap" of the material can produce electricity in a solar cell.

Solar Electric (PV) panels have remarkable efficiencies. PV Panels, converting more sunlight into electricity, produce impressive amounts of power which you can tap. Solar PV Panels are approaching 14% conversion efficiencies in the field. Solar PV Panels can produce about 140 watts, per square meter of area, at peak.

Using Solar Energy to do work:

This eBook uses examples of solar energy to produce electricity. Solar electricity is used to charge a battery system. The solar-charged battery will power an inverter, to provide standard AC current which, in turn, powers your AC loads, such as Home Power.

Solar PV Power Systems, contain "Three" basic elements. Energy In, Energy Storage, and Energy Out.

Energy - In, will include your Solar PV panels, Racking Hardware, Site Preparation, System Assembly, and Wiring.

Energy - Storage, refers to your Charge-Controller, Battery System, Fusing and Safety Disconnects.

Energy - Out, determines the Output energy form you need to power your Electric Load DC, or AC, for example, Single Phase 120 VAC at 60 Hz (American Voltage), running AC Electronics requires an Inverter.

To use Solar Energy for powering electronics, the next chapter, discusses the Step by Step procedure for designing, and sizing your Solar PV Power System.

Chapter Two - Defining Step-by-Step the Best Solar Power System for your Job

Solar Power supplies are useful in remote locations, or where electricity is not available to power your electric load. This Book covers Solar PV power systems which are stand-alone power supplies for many electronics, and other remote site power supply requirements.

Power remote cameras, electronics, GPS, scientific sensors, LED lighting, water pumping, UV water treatment, Electronics, Sensors, Communications, and Off-Grid homes ,and facilities, can all be powered with Solar PV power plants.

The following Steps define your system, for choosing the right Solar PV power supply:

Step One: Can my load be powered directly by the sun, or do I need a battery bank for power on demand?

If your project is water pumping, for example, then your system can be powered directly with solar PV panels, through the pump controller, and you don't need batteries.

If your project requires "controlled" power such as an LED area sign lighting systems, cameras, Sensors, or Home Power, then you'll need batteries. The systems, included below as examples, will all use Batteries, as the most common application for remote sites is power on demand 24 hours per day.

If your "electric load" needs power "on-demand," then you'll need Batteries in your Solar PV system.

Step Two: Is my electric load DC, or AC electricity?"

Electric loads are either Direct Current (DC), or Alternating Current (AC). Your electric load may be a camera, LED lights, or other electronics which are DC in draw. (Plug in power supplies convert grid AC into DC).

If your load is DC, match the DC Voltage of your device, with the Battery Voltage, and run your load directly from the batteries. Solar PV panels, to charge the batteries, are "arrayed" to match the Battery voltage. This is your DC System Voltage, and is usually 12, 24, or 48 VDC.

Note: if your Device Voltage is less than 12 VDC, such as 4.5 VDC, then add a DC to DC Step Down converter between the Battery and the Load.

If your electric load requires AC, such as Home Power systems, or AC motors and Pumps, then connect an Inverter to the battery bank. The systems listed below will include AC options so you can size your inverter properly for your load.

Alternating Current (AC) comes in Two Voltage Types: American, and European.

Single Phase American Voltage uses 120 VAC at 60 Hz, while European Voltage uses 220 VAC, at 50 Hz. Note: Solar PV off grid systems on the DC side are the same. The inverter your connect "defines" the AC output. For European Voltage, and Current, a European Voltage Inverter must be designated.

Examples, below are based on American Voltage, unless otherwise specified.

American Voltages come in Three Types: Single Phase, 2 Phase (Split Single Phase), and 3 Phase, each with different voltages respectively.

Your "electric load" will determine whether you'll use Single Phase 120/240 VAC 60 Hz, or 2 phase 208/240 (Split Phase with Ground), or 3 Phase 240/440 AC power.

Solar PV Battery Power Systems can support all of these Inverter Types.

AC electricity travels as a sine wave. 2 Phase (Split Single Phase), and 3 Phase formats have multiple Sine Waves (legs) transmitted simultaneously, in different phases. Multi-phase (Polyphase invented by Tesla), AC power transmission pushes each of these "legs" in different Phases.

For example, a 3 Phase AC power transmission has three "legs," each, 120 degrees out of phase. The relationship between the "legs" is called the Power Factor.

Modern Inverters have advanced "Power Conditioning" protocols which provide steady, and clean "sine wave" outputs keeping your electronics safe, and powered.

Step Three: What is the "Power" required by my electric load?

Electric loads, from LED lighting, security lighting, sensors, to home power, can all be defined with a Power demand, or "Power Rating." The "power" required, is the draw your device, or devices, pulls on the battery bank when on. Typically, if everything were turned on, what would be the total Power draw?

To know your Power draw, add up all the individual power ratings of all devices which will be powered by the system. If you're powering ten (10) separate LED high power Sign lights, with a draw of 30 watts per light, the total power draw will be 300 watts.

To calculate your projects "power" demand, list all appliances, and electrical loads you'll be powering. For example, if your remote Cabin has a Microwave, TV, Lights, and Radio, then list each appliance in a column.

Step Four: What is the "Energy" required by my electric load?

Energy is the measure of Power over Time. If your Power load is 1,000 watts (1 Kw), then to power that load for one hour, requires One Kilowatt-hour of Energy. One kWh of Energy provides the Power of 1,000 watts continuously over one hour of Time.

To calculate the Energy your Solar PV system needs to generate Multiply your Total Power Demand with the Hours per Day that load will run.

Step Five: How much Solar Energy do I have on my Site?

The Sun is a powerful source of energy.

In terms of actual power, the sun is rated at Standard Test Conditions (STC).

Standard Test Conditions define the peak "power density" of solar energy at the surface of the Earth at 1,000 watts of power per Square Meter (about 10.5 square feet). Note: STC also defines the amount of air-mass the sun path travels through (1.5 AMO at a 45 degree angle), standard temperature of 25 degrees C. (77 degrees F.). A wind speed of 2 meters/second further defines a standard condition for testing, and rating solar PV panels.

To determine how much Solar Energy you have at your location, look up the **Sun Peak-Hours** for your location on a Solar Map. In our examples here we're using a location in Kansas, with 5.5 Solar Peak-hours. Look up your locations solar peak-hour rating.

Raw solar energy produces, at peak condition during a clear sky, 1 Kilowatt (1,000) watts of optical power. Solar electric modules (Photovoltaic PV Panels) convert this optical energy into Direct Current (DC) with good efficiency delivering about 140 watts of electricity per square meter. Solar PV panels are "hardwired" to produce a desired voltage.

Each solar "Cell" produces about 1/2 Volt DC on its own. Amazingly, even under cloudy conditions solar cells produce good voltages. The amount of solar energy will drive the amount of "Current" the solar cells produce.

More direct sun on your Solar PV panel, much more current is produced. Solar cells are interconnected to produce Solar modules which, wired in series, produce working voltages to do work.

One square meter of sunlight is a power electrical force. Solar PV Panels at 14% efficiency would produce 140 watts, at 12 VDC in one m2. One square meter of solar energy can deliver over 11 Amps of current. 1,000 watts/m2 is a respectable power density.

The Energy produced by your Solar PV array will be the Power Rating of the Panels multiplied by the Sun Peak-Hours for your location. Rated power times Peak-Hours/day gives you the Energy your Solar PV panel, or PV Array, is expected to deliver daily.

Step Six: How to Size your Solar PV power system

From the Chapters below, select the best solar PV power system for your project. Match your Energy

Demand, with the Energy Production to find the best system.

If you'd like to calculate your own System Design the conventions follow these Steps:

Step One: List All your Loads by Power Rating

Example:

TV - 50 Watts
Microwave - 300 Watts

Step Two: List the "Hours per Day" your Load Runs:

TV - 4 Hours
Microwave - 1 Hour

Multiply "Power Watts" by "Hours" to get Energy in "Watt-hours"

TV - 50 watts Times 4 Hours/day = 200 watt-hours
MIcrowave - 300 watts Times 1 Hour/day = 300 watt-hours

Step Three: Add all Daily "Energy per Day" Loads for a Total

TV - 200 watt-hours + MIcrowave -300 watt-hours

Total = 500 watt-hours per Day

Step Four: Divide your "Total Energy per Day", by your location's **Peak-Hour rating**

500 watt-hours/5.5 Solar Peak Hours (Location dependent) = 90 Watts

This calculation says 90 watts of solar PV panels will produce, on average, the equivalent of 5.5 peak hours of operation, per day. This produces (90x5.5), or approximately 500 watt-hours of energy per day.

Round trip efficiencies require we "derate" this figure. Theory, and Practice, are related, but not twins. Derate your PV Panels for losses through the batteries, and the inverter by 25%.

The 90 Watt solar PV panel, in this example, becomes 90 watts multiplied by 1.25 for an solar PV Panel rated at 112.5 watts. Since manufacturers, typically, don't make a 112.5 watt PV panel, round up to the next available size. In this example, a 120 Watt PV panel would be a good selection.

Note: Increase the size of your Solar PV array, and Battery size if in an extreme location (very cloudy for example).

Step Five: Calculate your Battery and Charge Controller

Working backward from your load, in this example, 500 watt-hours per day, we can size the other equipment. To calculate your battery size, first choose a System DC Voltage. In this example, we'll choose 12 VDC.

Note: the larger your Solar PV System, the higher the DC system voltage should be.

Dividing Energy by System DC Voltage, (500 watt-hours/12 VDC) we arrive at a Battery Capacity, approximately 40 Amp-hours, in this example.

"De-rate" the Battery System, by 15%, to account for natural variation, and our Battery Capacity (Amp-hour Rating) of 40 Ah is multiplied by 1.15 giving us an Amp-hour rating of 46 Ah.

Battery manufacturers, may not manufacture a 46 Ah battery (at 12 VDC), so round up to the next highest rating, of available batteries. In this example, MK sealed maintenance free battery rated at 50 Amp-hours.

Charge Controllers control your Battery Charging System.

Choose your Charge Controller based on the DC Amp input (coming from your solar PV panel), with your PV System Voltage matching your Battery

Voltage. In our example, our DC System voltage is 12 VDC. Our PV Panel is 120 watts at 12 VDC. The Amperage of the Solar PV panel is 10 Amps (120 watts/12 VDC).

Therefore, your Charge Controller, in this example, will be rated at 12 Amps at 12 VDC.

Step Six: Choose your Inverter

If your "electric load" requires AC, then you'll need an Inverter. Inverters convert your Battery DC Voltage to Single Phase, Split-Phase, or 3 Phase AC output depending on your choice. Your Electrical Load, again, determines the AC electricity you must provide. Base your Inverter choice on the Power Load you intend to run. In our example, our two appliances have a total power demand of 350 watts (50 watts plus 300 watts).

Choose your Inverter sized above your Power Load. In this case, we'll choose a 350 watt to 500 watt inverter.

Inverters are also selected based on DC Input Voltage. Inverter models specify 12, 24, or 48 VDC input. Select your Inverter DC input Voltage to Match your Battery DC Voltage.

Inverter AC output will be Single Phase 120 VAC at 60 Hz, unless otherwise specified.

Visit the **Quick Guide** in Chapter Nine to a List of Sample Solar PV Power Systems

Once you know these vital statistics about your solar PV power project your solar supplier can know how to configure your system. Match the systems presented in this ebook which most closely meet your Energy requirements. If you don't see a system powerful enough for your needs listed in this ebook, then please visit **Solardyne.com** for more information.

Chapter Three: Solar Power using Solar Photovoltaic (PV) Panels for Remote Power Supplies

The Sun is a powerful source of energy, and ideal for powering electric loads in remote sites in the field. Solar energy, has a power density of 1,000 watts per square meter, offering a robust power supply.

Solar Electric PV modules (panels) produce strong DC currents, and well suited to extreme locations for their proven durability, and reliability in the field, over decades. Solar PV panels produce strong voltages, even in low light levels giving you some ability to charge your battery bank even in cloudy weather. Solar PV arrays are configured to provide specified performance over a wide range of climate conditions.

Off-Grid Electrical Loads require a power supply. The total "Energy" required to power an electrical load is calculated by knowing the Power Demand, and the Hours per Day, you operate the equipment. Energy, equals Power over Time. Size your Solar PV Power system to deliver enough Energy to drive your load day to day.

The Solar Power System for Off-grid loads will include the Solar PV panel array, with the mounting hardware to attach, and deploy your PV panels onsite. The DC electricity from the Solar PV panels is connected to a Charge-Controller.

The Charge-Controller is the "**brain**" of the system, and performs several functions to keep your power system safe, and operating efficiently. The Charge controller adjusts the power coming from the Solar PV panel by finding it's Maximum Power Point. Controllers use this Maximum Power Point Tracking (MPPT) to match the ideal draw from the PV panels to charge the specific voltage of the batteries.

The Charge-controller, also monitors the battery working voltage, and provides protection for the battery from two conditions: High Voltage, and Low Voltage.

The High Voltage condition is when your batteries are beginning to over-charge. Over-charging is dangerous for batteries, and can lead to failure. The charge-controller senses this condition, and employes a High Voltage Disconnect (HVD).

The HVD tells the controller to open the circuit from the solar PV panels so that no more charging can occur.

On the other side, if the Battery Voltage is sensed by the controller to be too low, the controller uses a Low Voltage Disconnect (LVD) to turn-off the circuit to power the load, and no more power is drawn from the battery. The LVD condition, is also dangerous to batteries, and is used to further protect the circuit.

Vital electrical loads such as sensors, cameras, and UV water sterilizers, for example, require power production on demand 24/7.

To do this we use a battery bank to store energy from the PV panels and provide power for your electrical load. The battery bank examples, listed below in the sample systems, are based on the total Energy required by your load to run for a given number of hours per day.

Regarding Power Supplies, all voltages run "downhill." If you want to power a 12 VDC load from a solar PV panel, you'll need to produce more than 12 VDC in voltage to drive the load either from a solar panel or battery.

For a 12 VDC Solar PV panel to produce a higher voltage the manufacturer will wire 36 individual solar cells in series within the module. Wiring the individual solar cells in series "Adds" the voltages producing a nominal 18 VDC.

Under load, when you connect your electric load, the voltage will drop as the solar PV panels drives the system.

Smaller solar PV panels from 60 to 135 watts are usually 12 VDC Panels. If you want larger system voltages wire these panels in series. Two in series for 24 VDC. Four in series for 48 VDC. Larger solar PV panels, from 140 watts - 280 watts are wired at 24 VDC each. Wire two 24 Volt PV panels in series for 48 VDC systems.

The DC Voltage of the Solar PV system is matched to the Battery Voltage, and to the Inverter Input Voltage you choose to Power the Load. This is your System DC Voltage.

The Solar DC Voltage will match the Battery Voltage, which in turn, matches the Inverter DC Input Voltage.

Note: Wiring solar PV panels in Series to increase Voltage (current remains the same), wire in Parallel to increase Current (voltage remains the same).

The energy produced by your Solar PV panel will be the power rating multiplied by your Daily Solar peak-hour rating for your site.

Check yoru location with a Solar Power Map, and note how many Solar Peak-Hours of solar radiation your site receives.

Mounting Your Solar PV Panels on location - The Options

Solar panels can be mounted a variety of ways. These options include Pole mounting, Ground mounting, Roof mounting, Passive Tracking, and Active Tracking mounting.

Top of Pole
Side of Pole
Ground Mount A-Frame Adjustable
Roof Mount Attached
Roof Mount Ballast-type
Passive Tracking
Active Tracking

Fixed mounts keep the solar PV panel at a specific Tilt-angle and is adjustable. To increase the output of your Solar PV array you can adjust this angle seasonally to maximize solar exposure. All Solar mounts are mounted to face South when your site is in the Northern Hemisphere, (Note: point your panels North, if you're in the Southern Hemisphere).

PV panels for power systems need a sturdy and reliable mounting bracket. Solar PV panels can be Pole mounted, either on the Top-of-the-pole, as a masthead, or can be Side-Pole mounted. Side-Pole mounting hardware has a bracket along the bottom and top of the Solar PV panels.

Pole mounting is a great option because it keeps your panels above the ground minimizing ground effects such as increased dust. Also, wiring your panels, once they're mounted on the Mounting Hardware bracket, is easier to do as crawling under the solar PV panels (J-Boxes are on the Back of the Panel) is handy.

Pole mounting your solar PV panels also makes installation easier. Smaller Solar PV panels will mount on standard 1.5" Schedule #40 pipe. Site preparation involves auguring a hole, and setting your pole in cement and aggregate.

Larger Solar PV arrays, up to 2,000 watts with Top of Pole mounting, will mount on either 2.5" Schedule #40 pipe, 3.5", or 4.5" pipe for the largest arrays. The examples below will call out the specific diameter of your mounting pipe.

Robust and low cost, you can also Ground Mount your Solar panels. Ground Mounting is an A-Frame rack that allows you to Adjust your Tilt Angle. The general ideal angle for mounting your Solar PV panels is found by taking your Latitude angle of the site, and subtract 15 degrees. If your location has a latitude of 45 degrees, the proper tilt angle is 30 degrees as measured from horizontal.

Note: If your site is in a Tropical Location, or a very Cloudy location, the best tilt angle is no angle. Mount your panels flat. This will receive the most "Global" solar radiation, that is both direct, and indirect rays.

You can also mount your solar PV array on your roof, if your roof is near your battery bank.

Solar energy production increases if you're always pointing the solar PV panel toward the sun. Tracking hardware does this either in one axis - Morning through Night, or on two-axis (Altitude and Azimuth) which is most accurate.

Trackers are categorized in two types: passive, and active, respectively. Passive tracking such as with the Zomeworks gear has great robustness, and increases Solar PV panel output in energy about 25% on average.

Passive-type trackers use uneven heating of internal gasses to self-adjust the panels throughout the day, following the sun. In the morning, these trackers reset to the rising sun and repeat the cycle.

Solar PV power systems work best in direct sunlight. Following the sun's path, solar PV panels increase energy production - power production over time.

Active (powered) tracking using Wattsun Active Trackers increases the output of solar PV panels as much as 35%. Using servo motors, and a solar sensor, powered by a separate solar PV array, the Wattsun trackers extract the maximum energy out of your Solar PV array.

There is a cost increase for the hardware, but system performance increases dramatically.

If your site is very remote, go with no moving parts, such as Top-Pole mounting requiring no maintenance potential. If you have easy access to your site, or you're in a very small foot-print, active-tracking is a great option for boosting performance.

In the sample systems listed below we'll use two Solar PV panels as examples. For smaller Solar PV panels, rated at 12 VDC each, the Dasol panels of 30, 60, 90, and 135 watt power, respectively are cited. For larger Solar PV panels we'll use the REC line using the popular and widely available 250 Watt module (panel) rated at 24 VDC each.

The Batteries, chosen for the Sample system example Part-Lists below, are Sealed-type, leak-proof, and maintenance free.

Sealed Gel Batteries are designed to be rugged, and are reliable. These Batteries can operate in any orientation (upside down not recommended), are manufactured for durability, and ship well. Safe, leak-proof, and powerful, today's Gel cells are convenient to work with in the field.

All Solar PV battery charging systems will use the properly sized Charge-Controller, which further protects the Battery Bank for reliable, maintenance-free operation.

An inverter is added to convert the DC capacity of the batteries to AC single-phase electricity to power the UV water treatment system.

Installation and Site Considerations for your Solar PV Power Supply

Solar Power System are likely located some distance from your Battery/Inverter power system. Ideally, your batteries and power panel (charge controller/ inverter fusing and safety disconnects) should be mounted indoors if the temperature drops below 4 degrees C. (40 degrees F.)

The optimum temperature range for Battery Storage equipment is between 9 degrees C., and 29 degrees C. The solar PV power system can be mounted up to 200 feet from the location of the Battery Bank being charged.

Note: If your Solar PV panels need to sited more than 200 feet away from your Battery Bank, and power system, you can increase the Voltage of your Solar PV array to compensate for the Voltage loss through a longer length of wire. Bring your Solar PV electricity in by wire to your Battery bank where your Charge controller, batteries, and Inverter are located.

If your site is in a Very Hot location increasing your Solar Array voltage by adding another panel, or substring of panels, in series to increase the voltage of the PV string.

Remote sites are notorious for logistical difficulties. Often, there is no power, which is the point of this eBook - powering UV water sterilizers with Solar PV power. As such, the sensitive electronics of your solar power system will require protection.

Included, in the examples below, are all weather battery boxes, which protect your batteries from the weather, and other environmental externalities. Battery boxes come either insulated, or non-insulated. If you're in a colder climate, then uses insulated. In temperate climates choose non-insulated. If you're in a hot climate use insulated.

The Solar PV panels will be Top-of-Pole mounted (other options exist, such as Ground, Roof, and Tracking mounts), to mount the Solar PV array to a Masthead. The masthead hardware fits on top of a vertical steel pipe (from 1.5-4.5" in diameter, Schedule #40 pipe) sunk into the ground for mounting the PV Panels.

Larger Solar PV arrays can use Ground Mounts as a stable, and reliable platform as the footings can be secured, important in extreme locations.

Chapter Four: Solar PV Power Systems from 30 to120 Watts

In this chapter, we'll look at the Solar PV Power supplies needed to run electronics such as Cameras, LED Lighting, and other low power electronic devices. Match your load's Energy Demand (kWh/day), to the following list of Solar PV Power systems Energy Production (kWh/day) to find the best match. For solar resource data on your site check the Solar Peak Hour Map at the National Renewable Energy Laboratory at this Link.

The following Solar PV power systems are configured for Energy on-demand 24/7

System Example A:

Power Rating: 30 Watts - 12 VDC - Energy Rating for Location with 5.5 Sun Peak-Hours 120 Watt-hours per day Energy Production. Monthly Energy Production: 3.64 kWh per Month.

Parts List:

Solar Array:

One (1) Solar PV panels rated at 30 watts and 12 VDC each. 30 watts total array
Example: Dasol DS-A18-30, Size each: 27.2" x 13.8" x 1" One (1) Top-of-Pole Mounting Hardware for one 30 watt panels (12 VDC). Mounts on 1.5" Schedule #40 pipe

Battery/Charge Controller/Inverter:

One (1) Charge Controller: SunGuard 4, rated at 4 Amps @ 12 VDC One (1) Battery: 12 VDC Battery, sealed, maintenance free MK Model 8G22NF rated at 40 Ah capacity.

Wiring, and site preparation is site-specific. Solar PV System DC Output: 12 VDC

If your load requires AC, add one of the following inverters:

Samlex: PST-15S-12A rated at 150 Watts
Cobra: 300 Watts
Morningstar SureSine: 300 Watts
Samlex PST-30012 rated at 300 Watts
Magnum: MM612 rated at 600 Watts
Samlex: PST-600 rated at 600 Watts

System Example B:

Power Rating: 60 Watts - 12 VDC - Energy Rating for
Location with 5. Sun Peak-Hours, is 240 Watt-hours
per day. Monthly Energy Production: 9.8 kWh per
Month.

Solar PV Array:

One (1) Solar PV panel rated at 60 watts at 12 VDC.
Example solar panel: Dasol DS-A18-60, Size each:
27.2" x 26.2" x 1.38"   One (1) Top-of-Pole Mounting
Hardware for one 60 watt panel. Mounts on 1.5"
Schedule #40 pipe.

Battery/Charge Controller/Inverter:

One (1) SunSaver-10, Charge-controller rated for 12
VDC battery charging up to 10 amps. One (1)
Sealed, Maintenance-Free Battery MK 8G22NF rated
at 12 VDC @ 50 Amp-hours. One (1) Side-of-Pole
Mounted Battery Box (mounted under the Solar PV
panels).

Wiring, and site preparation is site-specific. Solar PV
System DC Output: 12 VDC

Inverter Options include:

Samlex: PST-15S-12A rated at 150 Watts
Cobra: 300 Watts
Morningstar SureSine: 300 Watts
Samlex PST-30012 rated at 300 Watts
Magnum: MM612 rated at 600 Watts
Samlex: PST-600 rated at 600 Watts

System Example C:

Power Rating: 60 Watts - 24 VDC - Energy Rating for
Location with 5 Sun Peak-Hours 240 Watt-hours per
day Energy Production. Monthly Energy
Production: 9.8 kWh per Month.

Solar PV Array:

Two (2) Solar PV panel rated at 30 watts at 12 VDC
each wired in series for 24 VDC 60 Watt total array.
Example solar panel: Example: Dasol DS-A18-30,
Size each: 27.2" x 13.8" x 1" One (1) Top-of-Pole
Mounting Hardware for two 30 watt panels. Mounts
on 1.5" Schedule #40 pipe.

Battery/Charge Controller/Inverter:

One (1) SunSaver-10, Charge-controller rated for 24 VDC battery charging up to 10 amps. Two (2) Sealed, Maintenance-Free Battery MK 8G22NF rated at 12 VDC each @ 40 Amp-hours. One (1) Side-of-Pole Mounted Battery Box (mounted under the Solar PV panels).

Wiring, and site preparation is site-specific. Solar PV System DC Output: 24 VDC

Inverter Options include:

Samlex: PST-60024 rated at 600 Watts
Magnum: MM1524 rated at 1,500 Watts

System Example D:

Power Rating: 90 Watts - 12 VDC - Energy Rating for Location with 5.5 Sun Peak-Hours 370 Watt-hours per day Energy Production. Monthly Energy Production: 11.25 kWh per Month.

Solar PV Array:

One (1) Solar PV panel rated at 90 watts at 12 VDC Example: Dasol DSA18-90, Size each: 39" x 28.2" x 1.38" One (1) Top-of-Pole Mounting Hardware for one 90 watt panel. Mounts on 1.5" Schedule #40 pipe, augured into the ground with cement foundation

Battery/Charge-Controller/Inverter:

One (1) MorningStar Sunsaver 10, Charge-controller rated for 12 VDC battery charging. One (1) Sealed, Maintenance-Free Battery MK 8G24DT rated at 12 VDC @ 73 Amp-hours each. One (1) Chest Style Ground Mounted Battery Box (can be located up to 50 feet away from PV).

Inverter Options include:

Samlex: PST-15S-12A rated at 150 Watts
Cobra: 300 Watts
Morningstar SureSine: 300 Watts
Samlex PST-30012 rated at 300 Watts
Magnum: MM612 rated at 600 Watts
Samlex: PST-600 rated at 600 Watts

System Example E:

Power Rating: 120 Watts - 12 VDC - Energy Rating for Location with 5.5 Sun Peak-Hours 500 Watt-hours per day Energy Production. Monthly Energy Production: 15.2 kWh per Month.

Solar PV Array:

Two (2) Solar PV panel rated at 60 watts at 12 VDC, 120 Watts total wired in parallel. Example Solar module: Dasol DS-A18-60, Size each: 27.2" x 26.2" x 1.38" One (1) Top-of-Pole Mounting Hardware for two 60 watt panels. Mounts on 1.5" Schedule #40 pipe.

Battery/Charge Controller/Inverter:

One (1) Morning Star ProStar PS-15 MPPT, Charge-controller rated for 12 VDC battery charging up to 50 amps. One (1) Sealed, Maintenance-Free Battery MK 8G34 rated at 12 VDC @ 60 Amp-hours each. One (1) Side-of-Pole Mounted Battery Box (mounted under the Solar PV panels).

Inverter Options:

Samlex: PST-15S-12A rated at 150 Watts
Cobra: 300 Watts
Morningstar SureSine: 300 Watts
Samlex PST-30012 rated at 300 Watts
Magnum: MM612 rated at 600 Watts
Samlex: PST-600 rated at 600 Watts

System Example F:

Power Rating: 120 Watts - 24 VDC - Energy Rating for Location with 5.5 Sun Peak-Hours 500 Watt-hours per day Energy Production. Monthly Energy Production: 15.2 kWh per Month.

Solar PV Array:

Two (2) Solar PV panel rated at 60 watts at 12 VDC, 120 Watts total wired in series. Example Solar module: Dasol DS-A18-60, Size each: 27.2" x 26.2" x 1.38" One (1) Top-of-Pole Mounting Hardware for two 60 watt panels. Mounts on 1.5" Schedule #40 pipe.

Battery/Charge Controller/Inverter:

One (1) Morning Star PS-15 MPPT, Charge-controller rated for 24 VDC battery charging up to 15 amps. One (1) Sealed, Maintenance-Free Battery MK 8G34 rated at 12 VDC @ 60 Amp-hours each. One (1) Side-of-Pole Mounted Battery Box (mounted under the Solar PV panels).

Inverter Options:

Samlex: PST-60024 rated at 600 Watts
Magnum: MM1524 rated at 1,500 Watts

Chapter Five - Solar PV Power Systems 135 to 360 Watts

In this chapter, we'll look at solar PV power systems for Medium Range remote power. Area Lighting, communication stations, School trailers, remote Cabins can all be powered with solar PV power.

The Systems below use sealed, deep-cycle Batteries for safety, power capacity, and easy to use. Solar PV panels are mounted on Top-of-Pole racks, although you can always substitute other Racking, such as Roof, Ground, or Tracking. System DC Voltages will be either 12 VDC, or 24 VDC. All systems include Inverter Options.

System Example G:

Power Rating: 135 Watts - 12 VDC - Energy Rating for Location with 5.5 Sun Peak-Hours 550 Watt-hours per day. Monthly Energy Production: 16 kWh per Month.

Solar PV Array:

One (1) Solar PV panel rated at 135 watts at 12 VDC. Example PV panel: Dasol DS-A18-135, Size each: 27.2" x 26.2" x 1.38" One (1) Top-of-Pole Mounting Hardware for 135 watt panel. Mounts on 1.5" Schedule #40 pipe, augured into the ground with cement foundation.

Battery/Charge-Controller/Inverter:

One (1) Morning Star Prostar PS-15, Charge-controller rated for 12 VDC battery charging up to 15 amps. One (1) Sealed, Maintenance-Free Battery MK 8G34 rated at 12 VDC @ 60 Amp-hours each. One (1) Chest Style Ground Battery Box (can be located up to 50 feet away from PV).

Inverter Options include:

Samlex: PST-15S-12A rated at 150 Watts

Cobra: 300 Watts
Morningstar SureSine: 300 Watts
Samlex PST-30012 rated at 300 Watts
Magnum: MM612 rated at 600 Watts
Samlex: PST-600 rated at 600 Watts

System Example H:

Power Rating: 180 Watts - 12 VDC - Energy Rating for Location with 5.5 Sun Peak-Hours 740 Watt-hours per day Energy Production. Monthly Energy Production: 22 kWh per Month.

Solar PV Array:

Two (2) Solar PV panel rated at 90 watts at 12 VDC each wired in parallel for 12 VDC. Example: Dasol DSA18-90, Size each: 39" x 28.2" x 1.38" One (1) Top-of-Pole Mounting Hardware for two 90 watt panels. Mounts on 1.5" Schedule #40 pipe, augured into the ground with cement foundation

Battery/Charge-Controller/Inverter:

One (1) MorningStar ProStar PS-15, Charge-controller rated for 12 VDC battery charging rated at 15 Amps. Two (2) Sealed, Maintenance-Free Battery MK 8G22NF rated at 12 VDC @ 50 Amp-hours each. One (1) Chest Style Ground Mounted Battery Box (can be located up to 50 feet away from PV).

Inverter Options include:

Samlex: PST-15S-12A rated at 150 Watts
Cobra: 300 Watts
Morningstar SureSine: 300 Watts
Samlex PST-30012 rated at 300 Watts
Magnum: MM612 rated at 600 Watts
Samlex: PST-600 rated at 600 Watts

System Example I:

Power Rating: 180 Watts - 24 VDC - Energy Rating for Location with 5 Sun Peak-Hours 740 Watt-hours per day Energy Production. Monthly Energy Production: 22 kWh per Month.

Solar PV Array:

Two (2) Solar PV panel rated at 90 watts at 12 VDC each, wired in parallel for 12 VDC. Example: Dasol DSA18-90, Size each: 39" x 28.2" x 1.38" One (1) Top-of-Pole Mounting Hardware for two 90 watt panels. Mounts on 1.5" Schedule #40 pipe, augured into the ground with cement foundation

Battery/Charge-Controller/Inverter:

One (1) MorningStar ProStar PS-15, Charge-controller rated for 24 VDC battery charging rated at 15 Amps. Two (2) Sealed, Maintenance-Free Battery MK 8G34 rated at 12 VDC @ 60 Amp-hours each. One (1) Chest Style Ground Mounted Battery Box (can be located up to 50 feet away from PV).

Inverter Options:

Samlex: PST-60024 rated at 600 Watts
Magnum: MM1524 rated at 1,500 Watts

System Example J:

Power Rating: 250 Watts - 24 VDC - Energy Rating for Location with 5.5 Sun Peak-Hours 1,000 Watt-hours per day Energy Production. Monthly Energy Production: 30 kWh per Month.

Solar PV Array:

One (1) Solar PV panel rated at 250 watts at 24 VDC Example: REC Solar PV 250PE, Size each: 65.5" x 39" x 1.5" One (1) Top-of-Pole Mounting Hardware for two 250 watt panels. Mounts on 2.5" Schedule #40 pipe, augured into the ground with cement foundation

Battery/Charge-Controller/Inverter:

One (1) MorningStar ProStar PS-15, Charge-controller rated for 24 VDC battery charging. Two (2) Sealed, Maintenance-Free Battery MK 8G24DT rated at 12 VDC @ 73 Amp-hours each. One (1)

Chest Style Ground Mounted Battery Box (can be located up to 50 feet away from PV). One (1) ExcelTech XP/24 125 watt Single-Phase AC Inverter for 24 VDC.

Inverter Options include:

Samlex: PST-60024 rated at 600 Watts
Magnum: MM1524 rated at 1,500 Watts

System Example K:

Power Rating: 270 Watts - 12 VDC - Energy Rating for Location with 5.5 Sun Peak-Hours 1,110 Watt-hours per day Energy Production. Monthly Energy Production: 33 kWh per Month.

Solar PV Array:

Two (2) Solar PV panel rated at 135 watts at 12 VDC each wired in parallel, 270 Watt total array. Example PV panel: Dasol DS-A18-135, Size each: 27.2" x 26.2" x 1.38" One (1) Top-of-Pole Mounting Hardware for two 135 watt panels. Mounts on 1.5" Schedule #40 pipe, augured into the ground with cement foundation.

Battery/Charge-Controller/Inverter:

One (1) Morning Star ProStar PS-15, Charge-controller rated for 12 VDC battery charging up to 15 amps. One (1) Sealed, Maintenance-Free Battery MK 8G34 rated at 12 VDC @ 60 Amp-hours each. One (1) Chest Style Ground Battery Box (can be located up to 50 feet away from PV).

Inverter Options Include:

Samlex: PST-15S-12A rated at 150 Watts
Morningstar SureSine: 300 Watts
Samlex: PST-30012 rated at 300 Watts
Magnum: MM612 rated at 600 Watts
Samlex: PST-600 rated at 600 Watts
Magnum: MM1512 rated at 1,500 Watts

System Example L:

Power Rating: 270 Watts - 24 VDC - Energy Rating for Location with 5.5 Sun Peak-Hours 1,100 Watt-hours per day Energy Production. Monthly Energy Production: 33 kWh per Month.

Solar PV Array:

Two (2) Solar PV panel rated at 135 watts at 12 VDC each, wired in series for 24 VDC. 270 Watt total array. Example PV panel: Dasol DS-A18-135, Size each: 27.2" x 26.2" x 1.38" One (1) Top-of-Pole Mounting Hardware for two 135 watt panels.

Mounts on 1.5" Schedule #40 pipe, augured into the ground with cement foundation.

Battery/Charge-Controller/Inverter:

One (1) Morning Star ProStar PS-15, Charge-controller rated for 24 VDC battery charging up to 15 amps. Two (2) Sealed, Maintenance-Free Battery MK 8G34 rated at 12 VDC @ 60 Amp-hours each. One (1) Chest Style Ground Battery Box (can be located up to 50 feet away from PV).

Inverter Options Include:

Samlex: PST-60024 rated at 600 Watts
Magnum: MM1524 rated at 1,500 Watts
Magnum: RD1824 rated at 1,800 Watts
Magnum: RD2824 rated at 2,800 Watts

System Example M:

Power Rating: 360 Watts - 12 VDC - Energy Rating for Location with 5.5 Sun Peak-Hours 1,485 Watt-hours per day Energy Production. Monthly Energy Production: 45 kWh per Month.

Solar PV Array:

Four (4) Solar PV panel rated at 90 watts at 12 VDC each, wired in parallel for 12 VDC. Example: Dasol DSA18-90, Size each: 39" x 28.2" x 1.38" One (1) Top-of-Pole Mounting Hardware for four (4) 90 watt panels. Mounts on 2.5" Schedule #40 pipe, augured into the ground with cement foundation.

Battery/Charge-Controller/Inverter:

One (1) MorningStar TriStar TS-30, Charge-controller rated for 12 VDC battery charging. One (1) Sealed, Maintenance-Free Battery MK 8G24DT rated at 12 VDC @ 73 Amp-hours. One (1) Chest Style Ground Mounted Battery Box (can be located up to 50 feet away from PV).

Inverter Options include:

Samlex: PST-15S-12A rated at 150 Watts
Cobra: 300 Watt AC inverter
Morningstar SureSine: 300 Watts
Samlex PST-30012 rated at 300 Watts
Magnum: MM612 rated at 600 Watts
Samlex: PST-600 rated at 600 Watts
Samlex: PST-1000-12 rated at 1,000 Watts
Samlex: PST-1500-12 rated at 1,500 Watts

System Example N:

Power Rating: 360 Watts - 24 VDC - Energy Rating for Location with 5 Sun Peak-Hours 1,485 Watt-hours per day Energy Production. Monthly Energy Production: 45 kWh per Month.

Solar PV Array:

Four (4) Solar PV panel rated at 90 watts at 12 VDC each, wired two substrings of two panels in parallel, wire substrings in Series for 24 VDC Example: Dasol DSA18-90, Size each: 39" x 28.2" x 1.38" One (1) Top-of-Pole Mounting Hardware for four (4) 90 watt panels. Mounts on 2.5" Schedule #40 pipe, augured into the ground with cement foundation.

Battery/Charge-Controller/Inverter:

One (1) MorningStar TS-MTTP-45, Charge-controller rated for 24 VDC battery charging. Two (2) Sealed, Maintenance-Free Battery MK 8G34 rated at 12 VDC @ 60 Amp-hours each wired in series for 24 VDC. One (1) Chest Style Ground Mounted Battery Box (can be located up to 50 feet away from PV).

Inverter Options include:

Samlex: PST-60024 rated at 600 Watts
Magnum: MM1524 rated at 1,500 Watts
Magnum: RD1824 rated at 1,800 Watts
Magnum: RD2824 rated at 2,800 Watts

Chapter Six - Solar PV Power Systems from 500 to 1,500 Watts

As Solar PV Systems get larger, you'll see Voltages increase for the System DC Voltage. When electric current passes through wires, the resistance of that wire is to the Square of the Current. Double the current in a wire, and the Resistance to that current, in the wire, increases 4 fold.

To minimize "Current" losses, in a wire, we choose Higher Voltages. Power equals Voltage Times Amperage (P=VA). For a given power, say 1,000 watts, you can have that as 10 Amps time 100 Voltes (10x100=1,000).

Or, you can have 1,000 watts as 100 Amps, at 10 Volts (100x10=1,000). Both cases are 1,000 watts. However, the first case has a current rating of 10 Amps. The second case, has a current rating of 100 Amps. If resistance in a wire increases with the

Square of the Amperage, then we want to "minimize" amps, but still have high power. To do this, we convert to Higher, and Higher voltages, as power increases.

The following systems are for larger loads, such as water pumping on demand, LED Lighting systems for Sign Lighting, or Large Area Lighting, Remote Cabins, and Communication Stations.

System Example O:

Power Rating: 500 Watts - 12 VDC - Energy Rating for Location with 5.5 Sun Peak-Hours 2,000 Watt-hours per day Energy Production. Monthly Energy Production: 60 kWh per Month.

Solar PV Array:

Six (6) Solar PV panel rated at 90 watts at 12 VDC each, wired in parallel for 12 VDC. Example: Dasol DSA18-90, Size each: 39" x 28.2" x 1.38" One (1) Top-of-Pole Mounting Hardware for six (6) 90 watt panels. Mounts on 2.5" Schedule #40 pipe, augered into the ground with cement foundation.

Battery/Charge-Controller/Inverter:

One (1) MorningStar PS-45, Charge-controller rated for 12 VDC battery charging. Two (2) Sealed, Maintenance-Free Battery MK 8G24DT rated at 12

VDC @ 73 Amp-hours each wired in parallel. One (1) Chest Style Ground Mounted Battery Box (can be located up to 50 feet away from PV).

Inverter Options include:

Samlex: PST-15S-12A rated at 150 Watts
Cobra: 300 Watt AC inverter
Morningstar SureSine: 300 Watts
Samlex PST-30012 rated at 300 Watts
Magnum: MM612 rated at 600 Watts
Samlex: PST-600 rated at 600 Watts

System Example P:

Power Rating: 500 Watts - 24 VDC - Energy Rating for Location with 5.5 Sun Peak-Hours 2,000 Watt-hours per day Energy Production. Monthly Energy Production: 60 kWh per Month.

Solar PV Array:

Two (2) Solar PV panel rated at 250 watts at 24 VDC each wired in parallel, 500 Watt total array. Example: REC Solar PV 250PE, Size each: 65.5" x 39" x 1.5" One (1) Top-of-Pole Mounting Hardware for two 250 watt panels. Mounts on 2.5" Schedule #40 pipe, augured into the ground with cement foundation

Battery/Charge-Controller/Inverter:

One (1) MorningStar TS-MTTP-60, Charge-controller rated for 24 VDC battery charging. Two (2) Sealed, Maintenance-Free Battery MK 8G24DT rated at 12 VDC @ 73 Amp-hours each wired in series. One (1) Chest Style Ground Mounted Battery Box (can be located up to 50 feet away from PV).

Inverter Options:

Samlex: PST-60024 rated at 600 Watts
Magnum: MM1524 rated at 1,500 Watts
Magnum: RD1824 rated at 1,800 Watts
Magnum: RD2824 rated at 2,800 Watts

System Example Q:

Power Rating: 500 Watts - 48 VDC - Energy Rating for Location with 5.5 Sun Peak-Hours 2,000 Watt-hours per day Energy Production. Monthly Energy Production: 60 kWh per Month.

Solar PV Array:

Two (2) Solar PV panel rated at 250 watts at 24 VDC each, wired in series for 48 VDC, 500 Watt total array. Example: REC Solar PV 250PE, Size each: 65.5" x 39" x 1.5" One (1) Top-of-Pole Mounting Hardware for two 250 watt panels. Mounts on 2.5" Schedule #40 pipe, augured into the ground with cement foundation.

Battery/Charge-Controller/Inverter:

One (1) MorningStar TS-45, Charge-controller rated for 48 VDC battery charging. Four (4) Sealed, Maintenance-Free Battery MK 8G34 rated at 12 VDC @ 60 Amp-hours each wired in series for 48 VDC. One (1) Chest Style Ground Mounted Battery Box (can be located up to 50 feet away from PV).

Inverter Options:

OutBack GVFX3648 rated at 3,600 Watts

System Example R:

Power Rating: 1,000 Watts - 24 VDC - Energy Rating for Location with 5.5 Sun Peak-Hours 4.1 kWh per day Energy Production. Monthly Energy Production: 125 kWh per Month.

Solar PV Array:

Four (4) Solar PV panel rated at 250 watts at 24 VDC each wired in parallel, 1,000 Watt total array. Example: REC Solar PV 250PE, Size each: 65.5" x 39" x 1.5" One (1) Top-of-Pole Mounting Hardware for four 250 watt panels. Mounts on 3.5" Schedule #40 pipe, augured into the ground with cement foundation

Battery/Charge-Controller/Inverter:

One (1) MorningStar TS-MTTP-60, Charge-controller rated for 24 VDC battery charging. Two (2) Sealed, Maintenance-Free Battery MK 8G24DT rated at 12 VDC @ 73 Amp-hours each. One (1) Chest Style Ground Mounted Battery Box (can be located up to 50 feet away from PV). One (1) ExcelTech XP/24 125 watt Single-Phase AC Inverter for 24 VDC.

Inverter Options:

Samlex: PST-60024 rated at 600 Watts
Magnum: MM1524 rated at 1,500 Watts
Magnum: RD1824 rated at 1,800 Watts
Magnum: RD2824 rated at 2,800 Watts

System Example S:

Power Rating: 1,000 Watts - 48 VDC - Energy Rating for Location with 5.5 Sun Peak-Hours 4.1 kWh per day Energy Production. Monthly Energy Production: 125 kWh per Month.

Solar PV Array:

Four (4) Solar PV panel rated at 250 watts at 24 VDC each wire 2 panel substrings in parallel, wire those substring in series for 48 VDC, 1,000 Watt total array. Example: REC Solar PV 250PE, Size each: 65.5" x 39" x 1.5" One (1) Top-of-Pole Mounting Hardware for four 250 watt panels. Mounts on 3.5" Schedule #40 pipe, augured into the ground with cement foundation.

Battery/Charge-Controller/Inverter:

One (1) MorningStar TS-MTTP-60, Charge-controller rated for 24 VDC battery charging. Four (4) Sealed, Maintenance-Free Battery MK 8G24DT rated at 12 VDC @ 73 Amp-hours each, wired in series for 48 VDC. One (1) Chest Style Ground Mounted Battery Box (can be located up to 50 feet away from PV).

Inverter:

OutBack GVFX3648 rated at 3,600 Watts

System Example T:

Power Rating: 1,500 Watts - 24 VDC - Energy Rating for Location with 5.5 Sun Peak-Hours 6,100 Watt-hours per day Energy Production. Monthly Energy Production: 185 kWh per Month.

Solar PV Array:

Six (6) Solar PV panel rated at 250 watts at 24 VDC each, 1,500 Watt total array. Wire in Parallel. Example: REC Solar PV 250PE, Size each: 65.5" x 39" x 1.5" One (1) Top-of-Pole Mounting Hardware for six 250 watt panels. Mounts on 4.5" Schedule #40 pipe, augured into the ground with cement foundation.

Battery/Charge-Controller/Inverter:

One (1) MorningStar TS-MTTP-60, Charge-controller rated for 24 VDC battery charging. Four(4) Sealed, Maintenance-Free Battery MK 8G24DT rated at 12 VDC @ 73 Amp-hours each wired two substrings in parallel, substring in series for 24 VDC. One (1) Chest Style Ground Mounted Battery Box (can be located up to 50 feet away from PV).

Inverter Options:

Samlex: PST-60024 rated at 600 Watts
Magnum: MM1524 rated at 1,500 Watts
Magnum: RD1824 rated at 1,800 Watts
Magnum: RD2824 rated at 2,800 Watts
Magnum: RD3924 rated at 3,900 Watts

Chapter Seven - Solar PV Power Systems from 2,000 to 4,000 Watts

In this Chapter, let's examine Larger PV power systems. As you build larger Solar PV Power systems, pay close attention to your Grounding, Fusing, and Safety Disconnect. Power Panels include all power conditioning components all prewired, fused and assembled.

Large Solar PV power systems for remote Homes, and Businesses. These Sample systems, use deep cycle battery banks for Robust, and powerful energy delivery. Choose AC outputs for Single Phase, Split Phase, or 3 Phase power production.

Solar PV arrays, in these larger systems, will specify Ground Mounting, however, Pole Mounting is easy to work with, and gives a great look. As always, your choice of Mounting, or Racking hardware depends on the situation, and your preference.

Large Solar PV Systems:

System Example U:

Power Rating: 2,000 Watts - 24 VDC - Energy Rating for Location with 5.5 Sun Peak-Hours 8.2 kWh per day Energy Production. Monthly Energy Production: 250 kWh per Month.

Solar PV Array:

Eight (8) Solar PV panel rated at 250 watts at 24 VDC each, wired in parallel. 2,000 Watt total array. Example: REC Solar PV 250PE, Size each: 65.5" x 39" x 1.5" One (1) Top-of-Pole Mounting Hardware for eight 250 watt panels. Mounts on 5.5" Schedule #40 pipe, augured into the ground with cement foundation

Battery/Charge-Controller/Inverter:

One (1) AEE Power Panel OBJX5-GTFX3048 includes All fusing, disconnects, inverter, and charge controller up to 80 amps. Four (4) Sealed, Maintenance-Free Battery MK 8G24DT rated at 12 VDC @ 73 Amp-hours each. One (1) Chest Style Ground Mounted Battery Box (can be located up to

50 feet away from PV). One (1) Square D Safety Disconnect.

Inverter Output for AEE Power Panel: 2,500 Watts AC Single Phase 60 Hz

System Example V:

Power Rating: 2,000 Watts - 48 VDC - Energy Rating for Location with 5.5 Sun Peak-Hours 8.2 kWh per day Energy Production. Monthly Energy Production: 250 kWh per Month.

Solar PV Array:

Eight (8) Solar PV panel rated at 250 watts at 24 VDC each, wired 4 panel substring in parallel, those 2 substrings wired in series for 48 VDC. 2,000 Watt total array. Example: REC Solar PV 250PE, Size each: 65.5" x 39" x 1.5" One (1) Top-of-Pole Mounting Hardware for eight 250 watt panels. Mounts on 2.5" Schedule #40 pipe, augured into the ground with cement foundation

Battery/Charge-Controller/Inverter:

One (1) AEE Power Panel OBJX5-GTFX3048 with fusing, disconnects, Inverter, and Charge controller up to 80 Amps. Four (4) Sealed, Maintenance-Free Battery MK 8G24DT rated at 12 VDC @ 73 Amp-hours each. One (1) Chest Style Ground Mounted

Battery Box (can be located up to 50 feet away from PV). One (1) Square D Safety Disconnect.

Inverter Output: 3,000 Watts AC Single Phase 60 Hz

System Example W:

Power Rating: 3,000 Watts - 48 VDC - Energy Rating for Location with 5 Sun Peak-Hours 12 kWh per day Energy Production. Monthly Energy Production: 360 kWh per Month.

Solar PV Array:

Twelve (12) Solar PV panel rated at 250 watts at 24 VDC each, 500 Watt total array. Example: REC Solar PV 250PE, Size each: 65.5" x 39" x 1.5" One (1) Top-of-Pole Mounting Hardware for twelve 250 watt panels. Mounts on 6.5" Schedule #40 pipe, augured into the ground with cement foundation

Battery/Charge-Controller/Inverter:

One (1) AEE Power Panel OBJW5-GTFX3048Two (2) Sealed, Maintenance-Free Battery MK 8G24DT rated at 12 VDC @ 73 Amp-hours each. One (1) Chest Style Ground Mounted Battery Box (can be located up to 50 feet away from PV). One Square D Safety Disconnect. One (1) ExcelTech XP/24 125 watt Single-Phase AC Inverter for 24 VDC.

Inverter Output: 3,000 Watts AC peak at 6 Kw AC 120 VAC Single Phase

System Example X:

Power Rating: 4,000 Watts - 48 VDC - Energy Rating for Location with 5.5 Sun Peak-Hours 16.5 kWh per day Energy Production. Monthly Energy Production: 500 kWh per Month.

Solar PV Array:

Sixteen (16) Solar PV panel rated at 250 watts at 24 VDC each, 4,000 Watt total array wired 2 substrings in series for 48 VDC. Each substring of 8 panels in parallel. Example: REC Solar PV 250PE, Size each: 65.5" x 39" x 1.5" Two (2) Top-of-Pole Mounting Hardware for Eight 250 watt panels each, respectively. Mounts on two (2) 5.5" Schedule #40 pipe, augured into the ground with cement foundation.

Battery/Charge-Controller/Inverter:

One (1) AEE Power Panel OBJW5-GTFX3648. Power panel includes charge controller, fusing, disconnects and Inverter all prewired and tested, Eight (8) Sealed, Maintenance-Free Battery MK 8G24DT rated at 12 VDC @ 73 Amp-hours each. Two (2) Chest Style Ground Mounted Battery Box (can be located up to 50 feet away from PV).

Inverter Output: 3,600 Watts AC peak up to 7.2 Kw
Single Phase 120 VAC 60 Hz.

Chapter Eight: Quick-Guide to Solar PV Power Systems by Power and Energy Ratings

Listed, above in each chapter, are different Solar PV power systems to provide power, and energy on demand, for your remote site.

Solar PV Powered Energy Systems are designed to power electronics, to motors/compressors, to home power systems. Match the Energy Demand, of your site, or project, to the Energy Production of the following systems:

Systems are Rated by Power in Watts, System DC Voltage, and Energy Production per Day (based on 5.5 sun Peak-hours/day location)

System A - 30 Watts, 12 VDC, 120 watt-hours/day

System B - 60 Watts, 12 VDC, 240 watt-hours per/day

System C - 60 Watts, 24 VDC, 240 watt-hours per/day

System D - 90 Watts, 12 VDC, 370 watt-hours/day

System E - 120 Watts, 12 VDC, 500 watt-hours/day

System F - 120 Watts, 24 VDC, 500 watt-hours/day

System G - 135 Watts, 12 VDC, 550 watt-hours/day

System H - 180 Watts, 12 VDC, 740 watt-hours/day

System I - 180 Watts, 24 VDC, 740 watt-hours/day

System J - 250 Watts, 24 VDC, 1 kWh/day

System K - 270 Watts, 12 VDC, 1.1 kWh/day

System L - 270 Watts, 24 VDC, 1.1 kWh/day

System M - 360 Watts, 12 VDC, 1.48 kWh/day

System N - 360 Watts, 24 VDC, 1.48 kWh/day

System O - 500 Watts, 12 VDC, 2 kWh/day

System P - 500 Watts, 24 VDC, 2 kWh/day

System Q - 500 Watts, 48 VDC, 2 kWh/day

System R - 1,000 Watts, 24 VDC, 4.1 kWh/day

System S - 1,000 Watts, 48 VDC, 4.1 kWh/day

System T - 1,500 Watts, 24 VDC, 6.1 kWh/day

System U - 2,000 Watts, 24 VDC, 8.2 kWh/day

System V - 2,000 Watts, 48 VDC, 8.2 kWh/day

System W - 3,000 Watts, 48 VDC, 12 kWh/day

System X - 4,000 Watts, 48 VDC, 16.5 kWh/day

Click system links above for Specific Solar PV Power Systems. For Solar Resource Maps for Solar Peak Hours at your location click for NREL Solar Maps.

I hope you've enjoyed this ebook, and proves useful in planning your specific solar PV power project. For additional information on larger systems, and other clean energy topics please visit **Solardyne.com** on the worldwide web.

Enjoy your Solar PV Power Systems!